101

Red Hot

D&T
Starters

Louise T. Davies

Contents

Acknowledgements

I have been fortunate in working with so many enthusiastic and inspiring D&T colleagues over the years. In particular, I would like to thank the KS3 Strategy team (Jackie, Richard, Paul, Jayne and Jon) and the KS3 Pilot consultants. My thanks also go to the Royal College of Schools Technology Project team and the Teacher Fellows for their dedication and hard work, developing such great ideas for classroom teachers – many of those ideas influencing the activities in this book. And finally to Helen at Letts for believing that I did indeed have thousands of ideas, not just 101!

Introduction

Introducing starters

As endorsed by the Key Stage 3 Strategy Foundation Subjects: D&T, launched in 2004, lessons are divided into episodes of learning. There should be a clear start to the lesson, a series of activities to maintain pace throughout the middle part of the lesson and a plenary to draw together learning towards the end of the lesson. The starter can occur in many forms, from a written activity to a role play, with students working in groups, pairs or individually. This will largely depend on the size of your class and what you feel most comfortable with.

Target audience

This book is suitable for any D&T teacher who wants to be inspired, including supply teachers, cover teachers and NQTs (even primary teachers may find the book a good source of ideas which they can adapt). While targeted at Key Stage 3, the strategies for designing reflect ones that adult designers use, thus they can be used at Key Stage 4 and post-16 as well. While the activities use examples taken from food, textiles and product design, these examples can be changed, so that they can be used by any D&T teacher.

This book is for any teacher who wants to bring an enjoyable, positive atmosphere for learning into the classroom, with minimum time spent planning it.

In reality – the timing of starters and when to use them

The starters in this book are supposed to be flexible. They can take as little as five minutes if you feel this is enough to get your point across or can be used as a longer activity if you feel the students will benefit from that.

You may find it's best to start the lesson with a topic that is different from the main part of the lesson to allow students to warm up. On the other hand, it can be beneficial to start the lesson with an activity that is related to previous lessons, as it acts as a reminder and a revision tool. Students often forget what they did a week ago and it's a great way of refreshing memories. For practical lessons, short starters aim to get tools and equipment ready and remind students of safe processes, so that valuable time is not taken from their practical work. There are many starters to support designing in keeping with the Key Stage 3 Strategy aims to make designing fun, challenging and fast. These activities commonly use group work and discussion as this supports students generating, developing and evaluating ideas most effectively. Alternatively, these activities can be used when you feel the students are getting restless, or to restart the middle part of a longer lesson.

You may find some of the starters could be used as a plenary activity to test the students' understanding at the end of the lesson. This is a particularly good idea for the noisier activities, as the students leave your classroom excited rather than disrupting the main part of the lesson when you want them to settle to other tasks.

How to use this book

Contents grid

For simplicity, each starter has been assigned one specific year group and one objective from the Key Stage 3 D&T framework of objectives (for designing) or the Key Stage 3 programme of study (for making). However, all the activities are extremely flexible. Most of them can be used with any year group and adapted to your students' abilities and the project they are undertaking. In reality, the activities often cover more than one objective.

Aims and objectives

Each starter activity has an aim taken from one of the five strands of Key Stage 3 D&T objectives (exploring ideas and the task; generating ideas; developing and modelling ideas; planning; and evaluating) or the Key Stage 3 programme of study heading (making quality products). Each starter objective is then taken from the framework of objectives and the programme of study statements.

Resources

The activities generally require minimal resources. Most resources should be readily available in D&T classrooms, for example collections or photos of products that have been designed. However, some activities are so much more fun with interactive whiteboards, giant dice and cards. Once you have made up the cards, you will be able to use them over and over again. Photos and real products are inspiring when you are designing, though names of products on cards will suffice in some cases.

Activities

The instructions for each activity are presented as short, numbered action points, to minimise the time required for planning. Activities have been chosen that require no photocopying of additional material, except a few activities, which may benefit from packages, pictures or an OHT being prepared beforehand.

Hints

Useful tips are supplied to support the activity.

Follow-up

Ideas are given to help you link to the main activities of the lesson and where the starter might lead to or be extended to. Some useful websites are given.

Differentiation

Suggestions for adapting the activity are given where specific information is useful, but most of the activities are easy to differentiate. It is easy to give some groups more challenging questions or products and it is easy to mix up the groups so that different abilities work together or work in similar ability groups.

D&T teachers face two challenges: providing short starter activities at the beginning of a practical lesson when students cannot wait to get making, and also providing fun and engaging starters in lessons where students are disappointed that they are 'only designing'. I hope that this book is a collection of activities that gives students such a good start making and designing that they will ask enthusiastically, 'Are we designing today?'

Enjoy!

Healthy eating bingo

Aim
Exploring ideas and the task

Objective covered
To explore needs, wants and opportunities in the context of designing for themselves.

Activity

1 Copy this 3 × 3 grid onto the board:

Low fat	Low salt	High fibre
Fruit	Vegetables	Low sugar
High in calcium	High in vitamin C	High in iron

2 Ask students to copy a blank 3 × 3 grid onto a sheet of paper and put one food in each box that relates to the heading; for example, in 'low fat' the student could write 'baked beans'.

3 Ask each student in turn to name one food they have eaten today or yesterday, starting with breakfast, break, then lunch, tea, snacks and dinner.

4 If the name of the food is called out, they can cross it out on their grid.

5 First one with all the squares calls 'bingo'.

Follow-up
This leads into discussion of the healthy eating guidelines and the Balance of Good Health plate model. For more information, go to:
www.nutrition.org.uk/ balance.htm
www.wiredforhealth.gov.uk/home.php
www.food.gov.uk/healthiereating/

Hint
There are no such things as good and bad, healthy or unhealthy foods, only good and bad diets.

Differentiation
⊛ Introduce other nutrients, e.g. vitamins and minerals.

A...b...c... products

Objective covered
Exploring ideas and the task

Aim
To use existing, familiar products and systems to inform their design thinking.

Activity

1   As the register is called, students must answer with a product that begins with the same letter as their name (you can use first or family names).

2   Students have to listen actively so that they do not repeat products already called.

Hint
Students can say 'pass' and be asked to stand, returning to those students at the end of the register, once they have had some thinking time.

Follow-up
As a further memory test, ask the students to write down as many products as they can remember being called out. Then group them according to their age-related target market, for example, older people, children, teenagers.

Change its name

Aim
Exploring ideas and the task

Objective covered
To identify design possibilities by discussing needs and opportunities presented by the task.

Resources
Three or four product cards for each group – night light/lantern, game or puzzle, alarm, layered dessert, T-shirt, trainers, mobile phone.

Activity
✸ This is a helpful activity to use as you introduce the design brief.

1   Ask the students to work in groups of four to six.

2   Give them three or four cards with pictures or names of products.

3   Ask them to rename the products or items shown on the cards. Explain one example might be for a bike – you could call it 'a human-powered vehicle'.

4   Ask for feedback from the groups and explore whether this has opened up new ideas for them.

Follow-up
Calling a bike 'a human-powered vehicle' opens up new possibilities for designing and makes you think differently about the shape, size, configuration and methods of propulsion.

Look to the past

Aim
Exploring ideas and the task

Objective covered
To use existing, familiar products and systems to inform their design thinking.

Resources
For each group: a set of cards to show how one product has changed over time, e.g. telephones, chairs, computers, processed foods, ready meals, orange drinks, music systems, mobile phones, trousers/hats/ sportswear.

Activity
⚙ This is a helpful activity to use as you introduce the design brief.

1   Ask the students to work in groups of three or four.

2   Give them a set of cards that show images of how one consumer product has changed over time.

3   Get them to put the cards in order of time and identify what has changed (shape, colour, materials, how it is made etc.).

4   Take feedback from the groups. Ask them how the products reflected the fashions of the time.

Follow-up
To be creative we must first familiarise ourselves with the ideas of others. Ask the students to think about their own design brief – what else is like this? What could I copy? What idea could I incorporate? What other process could be adapted?

Sell, sell, sell

Aim
Exploring ideas and the task

Objective covered
To use existing, familiar products and systems to inform their design thinking.

Resources
Have props if you can (coat, hat, telephone product).

Activity
⊛ This is a helpful activity to use as you introduce the design brief to help students think about improving existing products.

1 Ask the students to work in groups of three or four.
2 Ask them to prepare a role play.

- One person is the customer complaining about how badly an existing item works (see Hint).
- One person receives a customer complaint and describes to the manager on the phone all the problems they have with this product.
- One person is the salesperson selling the new re-vamped version, solving all the design faults.
- One person is the sceptical consumer, checking that it is better than the last version.

Hint
Good products to choose
Existing items could include a vacuum cleaner, mobile phone, toaster, remote control for TV

Follow-up
Encourage the students to use this activity to work on their own design.

- Write the problem from your point of view.
- Write the problem from the perspective of at least two other people who are close to the problem.
- Synthesise the different perspectives into one all-inclusive problem statement.

Dig deep

Aim
Exploring ideas and the task

Objective covered
To find and select information that informs and clarifies thinking about the task.

Activity

1 Ask the students to work individually to start with.

2 Ask each student to use a large sheet of paper and write the product/context in the middle.

3 Explain that they are going to map all the information they currently know about the topic, by using key words linked to the theme in the middle. Ask them to draw quickly without pausing or editing.

4 Ask the students to look for relationships on their map – use lines, colours, arrows, branches or some other way of showing connections between the ideas. Using their own symbols makes the map visual which assists in their recall and understanding.

5 Ask them to discuss with a partner three things on their map that will help them think about the design and where they will need to go to find out more information.

Hint
Sketching a mind map or concept map encourages the 'tipping out' of information. It involves writing down a central theme and ideas which radiate out from the centre. By focusing on key ideas written down in the students' own words, and then looking for branches out and connections between the ideas, they map knowledge in a way that helps them remember new information and explore new ideas. This promotes non-linear creativity. This is a useful website: www.maps.jcu.edu.au/netshare/learn/mindmap/howto.html

Washing line

Aim
Exploring ideas and the task

Objective covered
To use solutions to problems from the present, and other times and cultures, to inform their design thinking.

Resources
Prepare cards with pictures or names of products. It is helpful to have enough for the whole class (or a group of six to eight).

Activity
1 Give each student a card (or pick a group of six to eight).

2 Ask students to peg their card on a 'washing line' in order of the oldest to newest product.

3 Discuss as a class and decide what you think is the right order.

Hint
Place key dates on the washing line. Useful website to get information for this activity:
www.ideafinder.com/history/timeline.htm

Follow-up
Ask students how they came to their decisions. What clues did they look for and what did they know about design trends and history that helped them?

Differentiation
⊛ Give the students the opportunity to ask one question of the teacher, in order to move their choice along the timeline or confirm their decision. Include some newer products that are made to look older as part of a fashion trend to promote discussion.

It's all part of the problem

Aim
Exploring ideas and the task

Objective covered
To explore and play with conventional and unconventional ideas related to the task.

Activity

⊛ This is a helpful activity to introduce the design brief.

1. Ask the students to work in groups of four to six and give them a set of statements related to their current design brief or project.

2. Explain that it is helpful to break down the problem statement, brief or theme into its component parts and to consider each in turn. Take the example statement: (Design and make) a **safety garment** or **accessory** for people aged 16 and under to **wear** when **walking, jogging** or **cycling** on the **road**.

   The key parts of this statement can be highlighted and questions asked about each part:

   - **Safety** – What will be the danger? What's a risk?
   - **Garment** – What sort? – hat/coat/band? Is it for everyone under 16? How big?
   - **Accessory** – What sort? How heavy? How will it be used? How will it stay on?
   - **Wear** – Pull on? Do up? Attach? Undo? Store? Look?
   - **Walking/jogging** – What time of day? Where will they be going? Why are they on foot?
   - **Cycling** – How fast? What sort of bike? How many bikes?
   - **Road** – What sort of road? How big? What traffic? Where on the road? On the pavement?

3. Give the students five minutes to brainstorm the parts of their statement.

4. Take short feedback from each group.

Who am I?

Aim
Exploring ideas and the task

Objective covered
To explore needs, wants and opportunities in the context of designing for clients.

Resources
Use a camera to take a series of photos hourly throughout a day that depict scenes from a specific person's life, without the person in them.

Activity

⊛ This activity shows how the design brief focuses on the needs of others. It helps the student to think beyond themselves.

1   Show the students the set of photos, to build up a picture of the day, marking them with the time.

2   Ask the students to work in groups of four to six to discuss the photographs and the clues within them about that person. Ask them to record notes using headings such as: Are they male/female? What is their family like? What do they do for work? Where do they live? What are their hobbies? What do they like?

3   Take feedback and clarify how the students came to those views, picking up any assumptions or stereotypes.

Follow-up
This information can then be used to design products for that person, based on their interests, hobbies, lifestyle etc.

Play with words

Aim
Exploring ideas and the task

Objective covered
To explore and play with conventional and unconventional ideas related to the task.

Activity
This is a helpful activity to use as you introduce the design brief to help the students examine assumptions.

1   Write a statement or design brief on the board. As a class, ask the students to identify the key words; highlight or underline them as they do this.

    Possible statements

    - With concern for the environment, there is greater demand for carrying devices to use when travelling on foot or on public transport.

    - An attractive, original and moving point-of-sale display is to be placed somewhere in a shop or supermarket.

    - Design a new and appetising filling for a pasty, which a target group of customers will want to buy.

2   Explain that it is helpful to think up alternative verbs and adjectives for key words. These deliberately change the way they look at the problem. For example, 'Draw' – alternative words might be sketch, attract, inhale, pull out.

3   Ask the students to work in groups of four to six. Give them five minutes to brainstorm alternative verbs and adjectives for the key words in the statement.

4   Take short feedback from each group.

Same product, similar needs

Aim
Exploring ideas and the task

Objective covered
To identify conflicting criteria and determine which should take priority.

Resources
A set of four to six products or photos of products. These products should be the same type of product designed for similar needs, e.g. a set of different torches (bike lamp, mini-torch, head-torch, freestanding torch), a range of yogurts (bio, children's, luxury, themed character/event), a range of purses/wallets (traveller's wallet, evening purse, themed character purse).

Activity
⊛ This is a helpful activity to use as you introduce the design brief.

1 Ask the students to work in groups of four.

2 Give the students a set of four to six products or photos of products.

3 Ask the students to think about how each product would be used and then list the most important criteria the designer would have worked to for each one. For example, the bike lamp would be easy to put on and take off the bike, rechargeable, waterproof, easy to switch on etc.

4 Take feedback from the groups.

Follow-up
Ask the students to put the criteria in order of importance. Ask them to apply this when they are coming up with criteria for their own design.

In ten years' time...

Aim
Exploring ideas and the task

Objective covered
To explore ideas in ways that show an understanding of their impact on the future.

Resources
One contemporary product, e.g. a fashion shoe, a light, a computer monitor, a chair.

Activity

⊛ This is a helpful activity to use as you introduce the design brief to help students have more imaginative responses rather than those that exist already.

1 Ask the students to work in pairs.

2 Show the class a product, such as a contemporary shoe, a light, a computer monitor, a chair. Each pair is asked, 'What will this product look like in ten years' time?'

3 Give them five to eight minutes to list five key points.

4 The pair shares their ideas with another pair. They have five minutes to discuss and agree six key factors from their combined list of ten. The group then agree the top three factors to report back to the class.

Hint
You can support this activity by having some factual 'headlines' of social and economic trends, e.g. increasing Internet use, growing number of older people, increased traffic etc.

Follow-up
Ask the students to think about how what they are designing now may change in the future, and if they can think about what future needs might be.

Why put it right?

Aim
Exploring ideas and the task

Objective covered
To discuss, debate, question and challenge information and the nature of the task itself.

Resources
One card for each group with photo/name of problem product.

Activity

⊛ This is a helpful activity to use as you introduce the design brief, to help them to understand why the project has been chosen.

1 Tell the class some stories or show them some real products that do not work well.

2 Ask the class to describe a product that they have used that does not work well

3 Ask the students to work in groups of three or four. Give each group a photo/name of a product and a serious problem (e.g. life-threatening). For example:

   • Rucksack – straps could get caught
   • Washing-up liquid bottle – children might be attracted to drink this
   • Airport luggage trolley or supermarket trolley – can tip over

4 Ask them to discuss ways that the product could be improved and why it is important to improve that product.

Hint
It is possible to focus on safety issues, aesthetics, function/user-friendliness, cost and environmental impact. Some interesting human factors case studies can be found on: www.baddesigns.com

Understanding individuals

Aim
Exploring ideas and the task

Objective covered
To recognise critical factors that need to be included as design criteria.

Resources
For each group: six product photos and six cards with photos/names of people.

Activity
⚙ This is a helpful activity to use as you introduce the design brief to understand designing for other people and specific needs.

1. Ask the students to work in groups of three or four.

2. Give each group sets of six product photos to look at (e.g. six different drinking bottles, magazines, games, music players, snack foods, beachwear).

3. Place photo/name of six different people who might use those products on the board (e.g. baby, toddler, school child, teenager, student, office worker, fitness fanatic, retired person).

4. Ask the students to discuss the needs of the people in the photos and to match them to the product (they can place them under the picture of the person on the board).

5. As a class, take feedback on how they decided which product matched the person and what gave them information/clues to help their matching.

Differentiation
⚙ This task is easier with one set of products, though you can build up interesting profiles if you ask the students to do this with a number of products.

You're a diamond!

Aim
Exploring ideas and the task

Objective covered
To identify conflicting criteria and determine which should take priority.

Resources
Nine cards with photos/names of products per group. For example:

- Most to least needed by society/useful
- Most to least liked by the group
- Most to least expensive
- Most to least damaging to the environment.

Activity

⊛ This is a helpful activity to use as you examine the design brief and the context of the user to establish what are the important criteria.

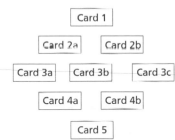

1 Ask the students to work in groups of three or four.

2 Give the students a range of nine cards and ask them to rank them into a diamond shape in order of importance.

3 Take feedback from the groups.

Hint
Diamond ranking can be used as a lesson starter in a number of ways. See page 71 for more information about ranking.

Line-ups

Aim
Exploring ideas and the task

Objective covered
To select information sources, gathering and sorting data that will help with ideas for, and decisions about, the design.

Resources
Six cards for each group of six with photos/names of products on.

Activity

1   Ask the students to work in groups of six.

2   Each student has a card and they line up in order (depending on the question set). For example:

   • Six photos/names of fruit (e.g. strawberries, lemons, oranges, apples, blackcurrants, melons)

      – Most juicy to least juicy fruit

      – Most vitamin C per 100g to least vitamin C per 100g

      – Cheapest to most expensive fruit

   • Six photos/names of fabrics/clothes (e.g. polyester duvet cover, acrylic jumper, PET fleece, cotton socks, wool blanket, silk tie)

      – Coolest to warmest fabric

      – Most wear-resistant to least wear-resistant

      – Cheapest to most expensive

   • Six photos/names of plastics (e.g. acrylic box, GRP boat, PET drinks bottle, nylon rope, polythene washing-up bowl, PVC hosepipe)

      – Stiffest to most flexible plastic

      – Lightest to heaviest plastic

      – Cheapest to most expensive plastic

3   Allow discussion in the group.

4   Review the line-ups and take feedback, agreeing the 'correct' order as a class.

Make a difference

Aim
Exploring ideas and the task

Objective covered
To explore ideas in ways that show an understanding of their impact on the future.

Resources
Six to eight pictures/names of products on A4 cards (e.g. biscuit, remote control, Lycra, game, medical kit, aircraft, bicycle, personal alarm, lantern, wind-up radio).

Activity
1   Ask six to eight students to come to the front of the class and work as a group to put the products in order of importance. Those that make the most difference to people's lives come first.
2   Allow the students to discuss the products and their importance to life.
3   Conclude the discussions reminding students that all the products that they design and make can and should make a difference to people's lives.

Hint
When ranking items, students must first clarify the purpose for ranking. Once the purpose has been determined, the criteria for ranking must be established. Ranking is often based on individual values and judgements about the relative importance of the criteria used in determining the rank of specific items. Therefore, it is important for students to give reasons for the rank orders they have selected. Verbalising their rankings helps students think about and clarify their choices.

Differentiation
As students complete the task, add further information to encourage them to reorder the products and challenge their assumptions about their value, for example, what if Lycra were used in a suit to enable a young person with cerebral palsy to move more independently?

Get in the hot seat

Aim
Exploring ideas and the task

Objective covered
To explore needs, wants and opportunities in the context of designing for markets.

Resources
Three to four cards with the name of an inventor of a famous product and information about them and their work.

Possible inventors cards
- Aeroplane – Orville and Wilbur Wright, American. Built their first powered machine, Kitty Hawk, in 1903. Made history's first powered, sustained and controlled aeroplane flight.

- Jeans – In 1872, Levi Strauss worked with Jacob Davis, a Nevada tailor, to make trousers with metal rivets at pocket corners and the fly. The first jeans came in two styles, indigo blue and brown cotton. Unlike denim, the duck material never became soft and comfortable so it was eventually dropped from the line.

- Microwave oven – Shortly after World War II, Percy Spencer stopped momentarily in front of a magnetron, in a radar. The chocolate bar in his pocket melted. Then he held a bag of popcorn next to it and watched as the kernels popped. From this simple experiment, Spencer and Raytheon developed the microwave oven. The first microwave oven weighed 337 kg and stood 165 cm high.
www.ideafinder.com/history/of_inventors.htm
www.invent.org/hall_of_fame/1_3_0_induction_meet.asp

Activity
1   Ask three or four students to come to the front of the class to 'hot seat'.
2   Give each student a card with the name of an inventor of a famous product and information about them and their work.
3   The class has to pose one question at a time in an attempt to guess who the inventor is.

Put to other uses

Aim
Exploring ideas and the task

Objective covered
To speculate about and envisage both common and unusual possibilities presented by the task.

Resources
A simple set of products, e.g. a paper clip, clothes peg, CD case, wallet.

Activity

⊛ This is a helpful activity to use as you introduce the design brief about a familiar product, to help them think more widely about its uses and possibilities.

1. Explain that an inventor produced a glue that did not work well. Show the students 'sticky notes' and say that it wasn't until the inventor found a use for the note as a reliable and reusable bookmark that this 'glue' became a useful product for people.

2. Ask the students to work in groups of four to six.

3. Give them each a product (e.g. a paper clip, clothes peg, CD case, wallet).

4. Ask them to think of as many alternative uses for it as possible. These can be simply listed or sketched.

5. Allow eight to ten minutes for this activity and then take feedback from the groups.

Hint
Shorten feedback by asking the group to select their most unusual or unconventional idea. There are lots of other examples to use: Garrett Morgan's hair-straightening cream was invented by accident. He was trying to find a chemical that would prevent his sewing machine needle from overheating; when the chemical made a dog's fur stick straight up, he knew he'd found something else entirely.

Shop 'til you drop

Aim
Exploring ideas and the task

Objective covered
To explore needs, wants and opportunities in the context of designing for markets.

Resources
A set of cards with the shopper profiles, magazines, scrap materials and images, A3 paper/card, glue.

Shopper profiles

- 'Impulsive' – under 35 years old, female, single or married with young children, has spending money, buys on spur of the moment, first to try new products, prefers well-known brands, doesn't bother to look for the lowest price.

- 'Independents' – 35–44 years old, men and women, single or married without children, high income, technologically aware, not first to buy new products, prefer well-known brands, want value for money, don't believe retailers give impartial advice.

- 'Aspiring' – 18–24-year-old men, single, first to try new products/technology, lowest price is most important, like store credit, like shopping!

Activity
1   Ask the students to work in pairs or threes.
2   Explain that some high street stores categorise their customers into groups of shoppers based on their market research. You can give each group a copy of these profiles.
3   Give each group magazines, scrap materials and images to use to create a *client profile collage* for one of these categories of shopper, depicting their interests, lifestyle, needs and wants.

Follow-up
Ask the students to list products or collect adverts from magazines that are aimed at each group.

SCAMPER

Aim
Generating ideas

Objective covered
To make connections and see relationships between the form and function of existing products and possible design proposals.

Resources
SCAMPER questions on separate cards:

S – **Substitute**: What could be used instead?
C – **Combine**: What can be added?
A – **Adapt**: How can it be adjusted to suit a condition or purpose?
M – **Modify**: How can the colour, shape or form be changed?
    Magnify: How can it be made larger, stronger or thicker?
    Minimise: How can it be made smaller, lighter or shorter?
P – **Put to other use**: What else can it be used for other than the original intended purpose?
E – **Eliminate**: What can be removed or taken away from it?
R – **Reverse**: How can it be turned round?
    Rearrange: How can the pattern, order or layout be changed?

Activity
1. Explain that the acronym SCAMPER will help them generate more design ideas and that each letter asks them to think about new design ideas.
2. Put them into groups of 3 or 4 and give them one or more SCAMPER cards. Allow each group five minutes to use the card to generate new ideas.
3. Take feedback from the groups.

Hint
Give the students a limited number of cards. 'P' (put to other uses) and 'M' (modify, magnify or minimise) are often successful starting points.

What is this?

Aim
Generating ideas

Objective covered
To develop the capacity to build images in the mind's eye.

Resources
Six cards with product photos/names, e.g. night light/lantern, game, puzzle, toaster, T-shirt, trainers, mobile phone.

Activity
✻ This activity is useful when you are trying to get students to generate a wide range of initial ideas.

1 Ask six students to come to the front of the class.

2 Give each student a card with a product on it, e.g. night light/lantern, game or puzzle, toaster, T-shirt, trainers, mobile phone.

3 Ask them to mime the product in use.

4 Get the class to guess the product.

Differentiation
✻ Have a hidden link between all of the products and ask the students to guess the link at the end.

Follow-up
Ask the class to divide into groups of four to six and devise a short sketch that links the uses of four of those products together.

Get close up

Aim
Generating ideas

Objective covered
To use a range of materials to stimulate the imagination.

Resources
A close-up image of an unusual product, e.g. doll's house furniture, models, miniature items, exhibition pieces, undefined objects. The product may have been photographed in such a way that dimensions, function, materials and potential end user are not clear.

Activity

⊛ This activity is useful when you are trying to get students to generate a wide range of initial ideas.

1   Show the class an image of an unusual design product.

2   Ask the students to work in groups of three or four to identify five to eight sources that would provide more information about the product and would help to develop further design criteria and the design itself.

3   Take feedback from the groups about what they think the object is, how they would find out more about it and what sources would help them do that. Discuss the possible best sources of information.

Follow-up
Ask the students to use the close-up image to transform it into part of another product.

Brainstorming

Aim
Generating ideas

Objective covered
To use a range of strategies to produce, communicate, record initial ideas to assist self-reflection and to describe their ideas and thinking to others.

Activity

1   Explain to the class that it is a technique for generating a wide number of ideas related to your particular design project, without evaluating them. It can take them outside the familiar into the strange or unexpected. Remind the class of the brainstorming rules:

   ● Choose one person to take notes or record in a spider diagram.

   ● Make sure all ideas are recorded.

   ● Agree not to make fun of anyone's ideas.

   ● Nothing is too silly to be suggested.

   ● Add to each other's ideas.

   ● Set a time limit (say five minutes).

   ● Always listen to what others say.

   ● Try to give as many ideas as others.

2   State the problem/task (e.g. bread products) and record it at the centre of the brainstorm sheet/flip chart/whiteboard.

3   Give students a time limit – experience will show how much time you need – the larger the group, the longer the time.

Hint

Laughing is to be encouraged, criticism is not. As soon as students fear criticism of their ideas, they'll stop generating them. Also, ideas that first seem silly may prove to be very useful or may lead to other interesting ideas. For more information:
www.jpb.com/creative/brainstorming.html

Time of day

Aim
Generating ideas

Objective covered
To use strategies that generate a variety of design ideas quickly as a direct response to the design criteria.

Resources
A series of photos that capture different times of the day (e.g. daybreak, sunset, break-time at school).

Activity
⊛ Choose a product that the students are currently generating design ideas for, e.g. shorts, game, puzzle, bag or a snack food product.

1 Ask the students to work in groups of four.

2 Show the students a current image of the product, e.g. some denim shorts.

3 Then show the students a series of photos that capture different times of the day (e.g. daybreak, sunset, break-time at school etc.).

4 Ask the groups to sketch quickly and annotate ideas to show how they would adapt the product for the time of day, e.g. morning – gym (change fabric to have some Lycra), night-time (change fabric to add a pattern, flame retardant finish).

Follow-up
You could use the time of year as a theme.

Word association

Aim
Generating ideas

Objective covered
To use a range of materials to stimulate the imagination.

Resources
A story book.

Activity

⊛ This activity is useful when you are trying to get students to generate a wide range of initial ideas by using word association.

1   Read the brief or project that the students have to generate ideas about.

2   Choose a story book and tell the students how many pages there are.

3   Ask students to choose a number between the first and last page of the book.

4   Write down the first noun on each page chosen on the flip chart (e.g. dog, face, jumper, house, disk, sound, shoe).

5   When you have about eight to ten words, choose one and show the students how to carry out a word association exercise. Use the random words to quickly generate themes. For example, the word 'face' might make you think about features of a face:

   ● Smile – products to celebrate success; amusing products to make you smile or to cheer someone up

   ● Eyes – products with decorations looking like eyes

   ● Ugly – scary/amusing products aimed at younger children

Activity continued

6 Point out that not all the ideas generated lead anywhere, but it is important to explore and record them all. But some of these ideas could be used to generate ideas for designs for their product.

7 Ask students to carry out word association in pairs or threes, choosing a number of the words to increase range of themes and possible ideas.

8 Ask them to feed back on their best idea.

Hint

Show word association being done quickly, with humour and unusual associations, without analysing them.

Follow-up

Students can use this strategy when they are generating ideas or when they are stuck. It is an interesting homework to set! Look at:

http://labs.google.com/sets

http://www.onelook.com/

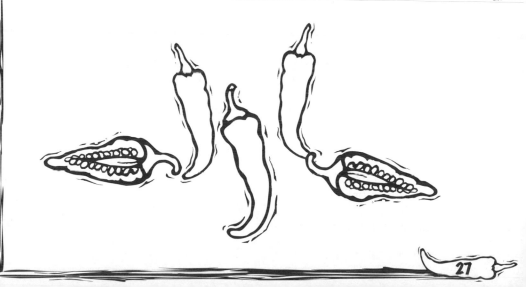

Extend and explore

Aim
Generating ideas

Objective covered
To make connections and see relationships between the form and function of existing products and possible design proposals.

Resources
A set of photographs or actual well-known branded products.

Activity

1 Use a photograph or real well-known branded product, e.g. Kit-kat, Fanta, Adidas football boots, iPod.

2 Put the students in pairs or threes and give them one product to focus on in their group. Ask them to identify the main features that give the product its identity, e.g. colour, flavour, texture, logo.

3 Give the students five minutes to come up with new ideas to extend that current range of products, e.g. Kit-kat ice-cream, Fanta sweets, Adidas tabletop football game, iPod watch.

Hint
The idea is to carry the unique features of the product into the new one, so that it is instantly recognisable as that product. If they struggle with this, ask them to imagine describing the product to someone who has never eaten/drunk/tried it before – what makes the product special and unique? You can ask students to keep within the current product range (i.e. food/drinks) or step beyond it into new sorts of products.

Follow-up
Ask students to describe examples of how one product has then extended into a product range and the different products that have been produced over a number of years. Ask them how these products have followed fashions and trends.

New from old

Aim
Generating ideas

Objective covered
To develop the capacity to build images in the mind's eye.

Resources
A set of photographs of existing products and set of names of new products, for example:

Existing product Starting with this . . .	New product . . . develop it into a . . .
rope	hat
jeans	bag
light-shade	storage container
CD	jewellery
pencil	clock

Activity
1   Ask the students to work in pairs or threes.

2   Give the students a set of photographs of existing products and then a random set of names of new products. How could you develop this rope into a hat?

3   Ask them to describe or sketch as many ways as possible to develop the existing product into the new product.

4   Take feedback on the most interesting ideas.

Follow-up
Encourage the students to think about how some new products have been developed out of old ones. Discuss why thinking about reusing and recycling are important and an interesting source of design ideas.

What would a famous person say?

Aim
Generating ideas

Objective covered
To use strategies that generate a variety of design ideas quickly as a direct response to the design criteria.

Resources
A set of cards with a famous person's name on each.

Activity

1 Remind the students of the design brief or introduce the project.

2 Ask them to work in pairs or threes.

3 Give each group a random card with a famous person's name on it. These can be sports personalities, TV celebrities, pop stars, royalty, charity promoters, business people, chefs, product designers, inventors. For example, if students are working on sports clothing, you may give them sports names to work with. If they are working on a vegetarian meal, you might give them chefs' names to work with.

4 Ask them to look at the name on the card.

5 Give the group five minutes to discuss and make a list of what the famous person would say about their problem.

6 Take feedback from the groups.

Follow-up
Discuss with the students why it is helpful to see the problem from another person's perspective. What would a person in a particular job say about your product?
http://gocreate.com/tools/jobs.htm

Put your finger on it

Aim
Generating ideas

Objective covered
To use strategies that generate a variety of design ideas quickly as a direct response to the design criteria.

Activity

⚙ This activity can take place to help students generate ideas for a project or to add another dimension when they have a few ideas.

1 Ask students to work as individuals or pairs.

2 Ask them to create a spider diagram with six 'legs'.

3 Ask them to choose words randomly from a dictionary and then add each word to one of the six legs.

4 Look at each leg and use the words to prompt ideas for new products.

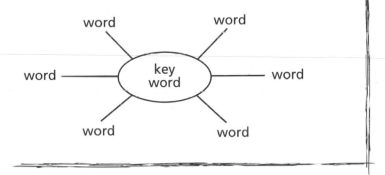

Hints
Using a dictionary rather than a random book does help, as students can check the meaning. This is like word association (see page 26).

It's a steal!

Aim
Generating ideas

Objective covered
To make connections and see relationships between the form and function of existing products and possible design proposals.

Resources
A series of products that focus on a specific user group, for example: baby carrier/sling, hard hat, stethoscope, emergency flare, staple gun, mobile phone hands-free headset.

Activity

⊛ This activity can help students generate creative new uses for a product.

1   Give some examples of how real designers have developed new ideas, by borrowing a solution or transferring it to a new application. For example:

   ● Lycra™ has been used in all sorts of stretch clothing. Now Lycra garments have been developed to support and comfort muscles and allow easy movement for people with cerebral palsy.

   ● James Dyson, borrowed the 'cyclonic' concept after seeing the vacuum extractor units that were used in a spray paint shop. He wondered if the idea might translate to vacuum cleaners.

2   Ask the students to work in groups of four to six and show them a series of cards with products on or parts of products. Ask them to think of imaginative new uses and new users. Who currently uses this? What is it useful for? Who else might find this useful? How else could it be used?

3   After ten minutes, take feedback from the groups.

Two-word technique

Aim
Generating ideas

Objective covered
To make connections and see relationships between the form and function of existing products and possible design proposals.

Activity

⚙ This activity is good when students are stuck with the design brief and becoming uncreative. Connecting words in unusual ways opens up new possibilities.

1   Explain that the meaning we give to certain words can block our ability to solve the problem.

2   Pick two words or phrases from the design brief. This should be a subject and an action verb. For example:

   • Chilled dessert and layer
   • Point of sale display and moving
   • Wallet and decorate

3   Draw a chart on the board and put one pair into the chart.

4   Ask the students to call out alternative words for them by using a thesaurus or brainstorming as a class.

Subject	Action verb	Sentences
Chilled dessert	layer	
Cold pudding	tier	
Frozen afters	coat	
Cool sweet	level	
Arctic	sandwich	

5   In pairs or threes, ask the students to try combinations of these words in sentences to give them some new ideas.

6   Take feedback on the sentences made up.

Hitch-hiking

YEAR 8

Aim
Generating ideas

Objective covered
To record and share their ideas with others and gather and use constructive feedback to develop a clear and detailed proposal.

Activity
⚙ This activity can follow brainstorming (page 24).

1   Ask the students to work in groups of three or four.

2   Each member of the group has a written list of ideas.

3   One person reads out their list. The others cross out any ideas that are also on their list and write down new ideas as they listen. This is called 'hitch-hiking'.

4   Each person reads out the ideas on their list that have not been crossed out and others cross out and hitch-hike as before.

5   Encourage the students to review the hitch-hiked list for new ideas. The best ideas may well come from the hitch-hiked list as the first ideas become developed and refined.

6   Take feedback from the groups on the interesting ideas that came out.

Hint
Encourage sharing of ideas, building on others' work, combining the best elements of all ideas. Some students may find this hard at first because in other subjects they would be discouraged from 'copying'.

Future gazing

Aim
Generating ideas

Objective covered
To develop the capacity for manipulating images of products in the mind's eye, in a constructive or analytical way.

Activity

This activity is useful when you are trying to get students to generate a wide range of initial ideas and not be restricted by today's products and technologies.

1   Start the lesson by giving the class a couple of statements:

- By 2020 half of the British adults will be aged over 50. Across Europe, some 130 million will be of that age.

- Around two million British people now work wholly or partially from home, a figure now expected to rise to more than 30 per cent of the entire UK workforce by 2006.

- Smart products which adapt to our own, personal requirements through use of embedded microchips will eventually replace mass market, one-size-fits-all products.

2   Ask the students to work in groups of four to six for five to eight minutes to discuss what these statements will mean to a product designer.

3   Take some feedback.

4   Ask the students to imagine what their home might be like in 25 years. Encourage them to be intuitive, spontaneous and outlandish. Allow them to sketch or describe their ideas.

Shopping list

Aim
Generating ideas

Objective covered
To develop flexible and independent thinking about existing products and solutions (by encouraging questioning, openness to ideas and approaches when generating design proposals).

Resources
Mail order catalogues.

Activity
✸ This activity is useful when you are trying to get students to generate a wide range of initial ideas by using existing products.

1 Ask the students to work in pairs and choose an item that they would like to buy from a shop in the near future.

2 Ask them to make a list of all the things that they consider important in choosing the product.

3 Ask them to sort the list in order with the most important point at the top of the list.

4 Take some feedback – when faced with the choices that are available in the high street, how do they decide which to buy?

Follow-up
Have some catalogues so that they can identify two or three products that appear to meet the points on their list. Ask them to use the list to choose the one they would buy. Ask them to present the list, product chosen and reasons for choosing in an interesting way.

Look to nature

Aim
Generating ideas

Objective covered
To produce and consider conventional, original, unusual, unique and/or eccentric ideas.

Activity

1   Explain that 'biomimetics' is about copying from nature. Designers get ideas from nature on many occasions. For example, Velcro was developed when George De Mestral's dog got covered in burrs from a bush after a walk. Smart textiles for extreme conditions were produced when Colin Dawson studied pine cones and the feather structure of Emperor penguins. A new swimsuit fabric which has less drag in the water was made by studying sharks. A new ultra sticky tape was developed by studying how geckos hang upside down by their hairy feet.

2   Ask the students to think of ideas that they can borrow from nature to help them design their product.

Hint

Possible questions depending on the design brief

*   Climbers need light and very strong ropes. Where might you look in nature for help with this problem?
*   Could you get ideas for a reflective safety garment by looking at animals or the natural world?

Follow-up
Discuss examples of how designers have used ideas from nature to solve their problems and how important it is for them to be open to new ideas, new attitudes and ways of doing things.

'Nothing new under the sun'?

Aim
Generating ideas

Objective covered
To use a range of design strategies which combine creative thinking and the application of knowledge.

Activity

⊛ This activity is useful when you are trying to get students to generate a wide range of initial ideas by looking at how they can adapt and develop new versions.

1 Design is about making existing ideas work better.

New possibilities for clothing

- Thousands die from skin cancer so researchers are developing clothes that protect us against harmful UV rays.
- Clothing is also being developed to change colour with the temperature to warn the wearer when they are in danger of hypothermia.
- Antibacterial fibres are now being used in hospitals.

2 T-shirts have been around since the 1920s. Ask the students to work in a group to come up with ideas on how to redesign a T-shirt for a particular person or situation with an imaginative new feature.

3 Ask the students to think about the product that they are designing and how they can make it work better.

Hint
Asking young people to redesign everyday objects can be unrewarding, particularly if students feel they are trying to compete with professional product designers. Ask them to design where there are no established products to copy.

Fat and skinny questions

Aim
Generating ideas

Objective covered
To draw upon a wide range of information sources including those not provided by the teacher.

Resources
A box with a concealed object inside.

Activity

❀ Fat and skinny questions is a strategy to help students think more thoroughly when gathering information about the project or idea. Fat questions require more thought, discussion. Skinny questions require a simple yes/no.

1   Conceal an object in a box and pass it around.

2   Draw a grid on the board. Students ask questions about the contents of the box and then rate the questions as either 'fat' or 'skinny' questions.

3   Record the students' questions on the grid, until students gather enough facts to know what is in the box. Open the box and reveal its contents.

4   Ask students to decide if questions are 'fat' or 'skinny' and why they think so. Which questions were most helpful in finding out what's in the box?

Question	Fat	Skinny

Follow-up
To provide practice for students in generating fat and skinny questions, ask them to create one fat and one skinny question about a product or picture.

Morphology matrix

Aim
Generating ideas

Objective covered
To record and share their ideas with others, and gather and use constructive feedback to develop clear and detailed design proposal.

Activity

1   Copy the table.

Name of part	Function of part	S	C	A	M	P	E	R

2   For the product they are working on, they have to break it down into its component parts and review one.

3   Put the name of all the parts in the first column.

4   Put a brief description of the function of the part in the second column.

5   Use the third column to write as many different ideas or solutions as they can think of (see SCAMPER, page 21).

6   Ask them to use a highlighter pen to mark up the most promising ideas in the third column.

7   Morphology is the process of selecting the best appropriate ideas for each part and combining them into a finished solution. Ask the students to look at the different combinations possible.

Take this theme and modify

Aim
Generating ideas

Objective covered
To record and share their ideas with others, and gather and use constructive feedback to develop a clear and detailed design proposal.

Resources
A selection of products, e.g. pair of shorts, night light, ready meal, storage container, puzzle.

Activity
1   Remind students of the current project or brief.

2   Draw a quick outline sketch of an existing product, similar to the one they are designing, on the board.

3   Unpick and describe the main features of the product, e.g. materials, cost, colour. Annotate those features on your sketch as you describe them, e.g. elasticated waistband, designed for children, pastel cotton, overlocked seam, single patch pocket, Velcro fastening, £4.99, machine washable.

4   Ask the students to work in groups of four to six.

5   Give each group a theme (see Hint) to work to and give them two minutes to think of how they can change the product to fit the theme.

Hint
Possible themes for groups
Summer, winter, healthy/sporty, Mexican (or other nationality), children's (or other age group), *Shrek* (or other film/fashion trend), delivery driver (or other occupation)

How might I...?

YEAR 9

Aim
Generating ideas

Objective covered
To produce creative solutions which address the design criteria in expected and/or unexpected ways.

Activity

1   Write on the board the following letters:

    SSKIXELTETCTHERS

2   Ask the students to copy them down in pairs.

3   State the problem as:

    'How can I cross out six letters to form a common word?'

4   Ask them to see if they can solve the problem.

5   Do not reveal the answer at this stage.

6   Write on the board the following letters:

    DSEIXSLIETGTENRS

7   Ask the students to copy them down in pairs.

8   This time state the problem as:

    'In what ways might I cross out six letters to form
    a common word?'

    Ask them to solve the problem.

9   The last phrase inspires the pairs to think of many alternatives, including the solution which is to literally cross out the letters S,I,X,L,E,T,T,E,R,S leaving the word DESIGN.

10 Now ask the students to rephrase their design problem, using the 'how might I...?' stem to see if this prompts new ideas for them.

42

Look from a different angle

Aim
Generating ideas

Objective covered
To analyse how existing products are designed and made, in order to provide a range of strategies and factual information to use when designing their own products.

Activity

1   Explain that it is important to look for new ways to tackle everyday problems. For example:

   • Nick Butler designed a radical new ironing board. The ironing board blows steam from the underside of the board to make your ironing easier. He used this idea after looking at industrial clothing presses. Effort is halved and clothing is perfectly ironed!

   • Alan Campbell, a 15-year-old student, designed a much faster-acting car brake signal. Alan looked at the problem in a new way. He looked carefully at what happens in an emergency. He noticed that your foot comes off the accelerator very quickly before hitting the brake. If the brake light sensed your foot suddenly coming off the accelerator instead of hitting the brake then seconds would be saved!

2   Ask students to discuss the scenarios in pairs.

   • Why do they think that designers had not changed the ironing board previously?

   • Why had car manufacturers not looked at what happens when we brake?

3   Ask them to identify a product that has remained the same design for a long time and to consider how it might be changed.

Follow-up
Look at a product used every day and think about new ways that could be developed to make the job easier.

Be your product

Aim
Generating ideas

Objective covered
To be prepared to take risks when generating ideas through a range of creative and critical thinking techniques.

Activity

⊛ This activity is useful when you are trying to get students to generate a wide range of initial ideas by understanding the product and context better.

1 Set the context for the design brief or product to be designed.

2 Start a story of a day in the life of the product (from the point of view of the product). Try to make the start adventurous and humorous, giving the product an 'emotional life'.

3 Ask the students to work in pairs to continue the story. Allow them about eight to ten minutes.

4 Choose a few of the students to tell their stories, perhaps by picking names at random from a hat.

Hint
Possible stories

- A bursting-at-the-seams gymkit bag travels to school.
- The TV set in a child's bedroom looks out on the world.
- The dog's lead gets used for walks, skipping and hanging out.
- The twins' pushchair goes shopping.

Follow-up
Ask the students to think of examples of products that we give characters to, e.g. a hedgehog boot scraper, names to our cars, mobile phone advertising, and whether this makes the product more appealing or user-friendly.

Get out of your box

Aim
Generating ideas

Objective covered
To produce creative solutions which address the design criteria in expected or unexpected ways.

Resources
Cards describing materials with special properties such as:

- Paper that lights up when you touch it
- Drink that digests harmful bacteria and cancer cells
- Fibre that is so strong when woven that it can never break
- Wood that is antibacterial with self-healing properties when it splits
- Metal like aluminium that is soft and bendy but will never break

Activity

1 Ask the students to work in a group of four to six. Explain that they are now working in the future, with new materials that do not exist in today's world.

2 Give each group a new material that has special properties.

3 Ask them to sketch two or three ways that this new material could be used in a product for the future. Explain that there are no physical limitations on what the material can do.

4 Take feedback from each group.

Follow-up
Encourage students to think of new applications for new products.

Polymorph – www.mutr.co.uk/prodDetail.aspx?prodID=576
QTC – www.peratech.co.uk
Softswitch – www.softswitch.co.uk
Nutraceuticals – www.nutrition.nestle.com/content.asp?THE_ID=127

Be whacky

Aim
Generating ideas

Objective covered
To combine ideas from a variety of sources.

Activity

1 Ask the students to work in groups of four to six.

2 Using a large sheet of paper, copy the chart below.

3 Choose a product or the context and put at the top of the sheet, e.g. the beach.

4 As a class, brainstorm lots of things that concern the subject and record them in the first column.

5 As a class, brainstorm words that describe the subject and record them in the second column.

6 Give the students five minutes to use coloured pens to connect words in both columns. Obscure connections often lead to the best ideas.

Product or context :	
Things that concern the subject	Words that describe the subject
e.g. sand, sea, cliff, swimming, rowing	e.g. wet, hot, humid, rocky

Hint
Record them as a list one beneath the other to make 'connecting' easier.

Follow-up
Affinity networks are a good way of grouping together similar or related ideas in a session where lots of ideas have been created. It can be done by grouping ideas using coloured pencils, making a new list with related ideas grouped together or drawing 'spider' diagrams.

Do the unexpected!

Aim
Developing and modelling ideas

Objective covered
To use a range of strategies to explore and experiment with ideas before making judgements.

Resources
A selection of real objects or photographs, e.g. trainers, bottle, drawing pins, CDs, toy, pencil, and seven word cards.

Activity

1   Ask students to work in pairs or threes. Give each group one object from a selection of real products/photographs or product names.

2   Each group is given a card with one of the following words:

Hide	Protect
Surround	Display
Support	Hold
Rest	

3   Ask each group to develop a design, using their word and the object they have been given. Give them five to seven minutes to develop the idea as a group.

4   Give each group ten minutes to model their idea, using materials available.

5   Take feedback from the groups.

47

Improve it

Aim
Developing and modelling ideas

Objective covered
To appraise ideas by continual reference to the design criteria.

Resources
A product close to what the class are designing, e.g. lasagne ready meal.

Activity

⊛ This activity can help students generate more creative ideas for a product because it breaks the product down into parts and asks them to look at each one.

1   Remind the class about the project and the design brief, and the design ideas they have initially generated.

2   Show them an existing product that is close to the one they are designing, e.g. lasagne.

3   As a class, list all the important features or attributes of the existing product and copy them into the first column (see ACCESS FM, page 76).

4   As a class, list all the characteristics in the second column.

5   Ask the students to work in pairs or threes to suggest what could be improved.

6   Take feedback and record these in the third column.

Activity continued

Product: Veggie lasagne ready meal		
Feature	Characteristics	What could be improved?
Sauce	runny tomato herby layers	
Pasta	plain wheat sheets	
Vegetables	carrots peas onions diced	
Topping	golden brown cream sauce	
Ready meal	baked microwaveable freezable single portion	
Nutrition	high salt high fat low fibre	

Follow-up

Ask the students to use a similar chart for their design ideas, listing the main criteria from the specification in the first column, the characteristics in the second and further developments in the third column. Discuss the third column with their group.

Model it

Aim
Developing and modelling ideas

Objective covered
To work out, and reflect on, the technical details of their ideas by modelling them through drawing, discussion, ICT and in 3D.

Resources
A selection of modelling materials and cards with ideas to model.

Activity
⚙ As part of the project, allow time for the students to 'play like a child', to experiment with ideas, images, materials, structures and use scrap materials to produce imaginative results.

1   Ask students to work in pairs or threes.

2   Give each group a small selection of modelling materials such as card, paper, scrap materials, mouldable materials.

3   Give each group a card with the name of the item you want them to model in 3D, e.g. CD rack, hat, wallet, torch, folding furniture, moving display, mask, puppet, computer mouse, bridge, dome, tent.

4   Get each group to show their design idea.

5   Ask the students how modelling in 3D helped them with that design.

Hint
Experimenting and playing are important for coming up with new ideas and ways of doing things. Sometimes these are found by accident! Encourage students to become less inhibited, and to express emotions and personal opinions.

Benchmarking

Aim
Developing and modelling ideas

Objective covered
To appraise ideas by continual reference to the design criteria.

Activity

This activity should take place when all students are developing one or two ideas and need suggestions that will help them work on their ideas further to match the specification more closely.

1 Draw a radar diagram or star chart on the board.

2 Label the chart with eight criteria, e.g. colourful, safe.

3 Discuss as a class and rate the criteria for the 'perfect' or ideal product.

4 Explain that this is their 'benchmark'.

5 Then ask the students to work in pairs. Take each idea under development and map them on the chart in a different colour:

- How closely do they match the 'perfect' product?
- What aspects of the design need to be improved?
- How can they get the design closer to the benchmark?

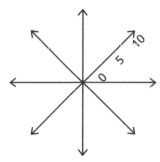

Dot voting

Aim
Developing and modelling ideas

Objective covered
To explore, experiment with and then select appropriate materials and processes.

Resources
Sheets of coloured dots.

Activity
This activity is helpful when all students are sorting out which ideas are worth thinking about further, e.g. after generating ideas activities (pages 21–33).

1 Ask the students to write down their ideas on cards or sticky notes.

2 Put all the ideas up on a flip chart sheet or board.

3 Allow students time to review all the ideas.

4 Give each student three sticky dots or votes. They put their dots on the three ideas that they think have most potential.

5 Discuss those ideas that get the most votes.

Hint
You can do this as a whole class or a smaller group. You may want to sort or categorise the ideas before voting and remove repetition.

Apply these words

YEAR 8

Aim
Developing and modelling ideas

Objective covered
To try fresh or alternative approaches when developing ideas and overcoming new problems and challenges.

Resources
Development ideas on cards, such as:

Multiply	Divide
Invert	Separate
Distort	Rotate
Flatten	Squeeze
Complement	Freeze
Soften	Fluff up
Add	Subtract
Widen	Repeat
Thicken	Stretch
Submerge	Protect

Activity
⊛ Use this activity when students have a few ideas generated but need help in developing them further.

1   Ask the students to work in pairs or threes.

2   Ask them to have their design sketches in front of them.

3   In turn they have to pick a card from the set on the tables and see if they can apply it to their design.

4   Ask for feedback from the groups on any further ideas they now have that they can develop.

53

Label it

Aim
Developing and modelling ideas

Objective covered
To develop different strategies to elaborate, embellish, expand and develop design ideas.

Resources
A list of brainstormed ideas on the board (see page 24) and labels for each group (e.g. excellent, likely, possible chance, 50-50, long shot).

Activity

❋ This activity should take place when all students have brainstormed a list of initial ideas and need to decide which ones are worth thinking about further.

1 Ask students to work in groups of four to six.

2 Read the list of possible ideas.

3 Ask the students to group them using the question, 'Will this product be successful?'

Excellent	Likely	Possible chance	50-50	Long shot
Will almost certainly succeed	Needs further refinement	Needs improvement	Could go either way	Remote chance of success

Follow-up
While you may not be interested in long shot or 50-50 ideas, the process of listing and grouping them expands your options. What would need to be done to move an idea from possible chance/likely to the excellent list?

Differentiation
❋ Instead of 'successful' you might introduce the term 'feasible and feasibility'.

Snowballing

YEAR 8

Aim
Developing and modelling ideas

Objective covered
To try fresh or alternative approaches when developing ideas and overcoming new problems and challenges.

Resources
A hat or box with names of all the students to pull out.

Activity
⚙ This activity is a good one for team projects where the team have a number of ideas that they want to combine and develop further.

1 Ask the students to work in a group of six.

2 Put all the names of the group on pieces of paper into a hat/box.

3 Each individual writes/sketches ideas.

4 Pull two names out of the hat – the two named students read one idea each out to the group.

5 The remaining group members try to integrate these two ideas into one idea.

6 Pull another name out of the hat – the named student reads one idea and the group members try to integrate it into the idea formed in step 6.

7 Put all names back in the hat – continue pulling one name out at a time, who reads an idea and the rest of the group tries to integrate it.

8 Continue until all the ideas have been read and integrated into one final solution. It may not be possible to integrate all ideas, but everyone gets a fair hearing.

Observing people and products

Aim
Developing and modelling ideas

Objective covered
To seek opinions of potential users of the product and decide whether their design criteria are accurate and detailed enough for the purpose.

Activity

1   Remind students that an important aspect of developing their designs will be thinking about how the user uses the product. Observing people using products helps the designers to understand needs better. For example, EEV were redesigning a thermal imaging camera for firefighters. The designers visited the crawling gallery that firefighters use in training. They found it terrifying to crawl through the smoke, and realised that the camera would often be used crawling on all-fours. This changed their thinking about the design.

2   Ask the students to discuss in pairs or threes how the designers might have changed the design once they had seen how the camera was to be used.

3   Ask them to draw a sketch of a camera that could be used by the firefighters to present to the class.

Follow-up
Encourage students to observe someone using an everyday product, such as a rucksack or mobile phone. Look for an opportunity to improve it.

Alternative questions

Aim
Developing and modelling ideas

Objective covered
To try fresh or alternative approaches when developing ideas and overcoming new problems and challenges.

Activity
⊛ This activity should take place when all students are developing one or two ideas further and need questions that will help them move forward.

1   Ask the students to work in pairs or threes. They should show their design ideas to the others in the group and explain them briefly.

2   Ask the group to copy this chart.

Design idea	Are there alternative materials, components or ingredients?	Are there cheaper materials of the same quality as higher priced ones?	Could the product be made lighter or smaller?	Are there any unnecessary parts?
Name.......... Idea............				
Name.......... Idea............				
Name.......... Idea............				

3   Ask them to spend eight to ten minutes discussing those questions as a group and helping each other record their answers.

4   Give the students three or four minutes to think about whether they could combine the best aspects into a new idea.

4 x 4

Aim
Developing and modelling ideas

Objective covered
To develop different strategies to elaborate, embellish, expand and develop design ideas.

Activity
1   Ask the students to work in groups of four or five.

2   Give each student a sheet of divided A3 paper.

1	2
3	4

3   Give the students four minutes to sketch, annotate or describe their design idea in the first section of their paper.

4   The students then pass their paper to someone else in the group. The next person has four minutes to add to the idea, drawing in the next blank section.

5   Repeat the four-minute slots until all sections are full.

6   The originator can then review the four developments and select or reject these suggestions.

Hint
You will need to give clear time reminders. Encourage students to:

● Focus on developing the idea, not coming up with new ideas.

● Add to the detail.

● Think about the materials.

What's it made from?

Aim
Developing and modelling ideas

Objective covered
To find out what materials and components are available and use technical information to decide on their suitability for the task.

Resources
Cards with names of products and features and properties of materials.

Possible cards

- Flat pack furniture (MDF) – I'm manufactured wood. I have a smooth surface and paint well. I'm found in most houses, especially student homes!
- Carved chair (mahogany) – I'm a dark colour and have a rich finish, I can be beautifully carved and stained even darker. You often see me in a dining room, where I often match the table.
- Zip (aluminium) – I'm used on many clothes to do them up. I am soft and bend easily, but I sometimes break.
- Remote control (polystyrene) – They mould me into shape even though I am stiff. I am light to hold. I make you lazy when we watch TV. You can drop me and glue me back together.

Activity

1   Choose six students from the class to stand at the front.

2   Give them a card with the name of a product (and features and properties of materials).

3   Each student in turn gives single clues about the uses and properties of the product until someone is able to guess the correct answer. Ask students to tell you the name of the product and the material it is made from, and what it was that gave the biggest clue. For example:

- I am used every day in the house.
- I have a lot to do with water.
- I'm very shiny when polished.
- You have to turn me to make me start.
  (Answer – Taps, brass)

Get a green card

YEAR 8

Aim
Developing and modelling ideas

Objective covered
To find out what materials and components are available and use technical information to decide on their suitability for the tasks.

Resources
For each group you will need 40–60 red, 40–60 green and 40–60 amber (yellow) cards. A set of statements, such as:

- You have reduced the amount of materials being used.
- You are making the product from materials from recycled products, such as glass, paper, aluminium foil or clothes.
- The materials you are using for your product will be recyclable.
- You have thought about what will happen when the product is thrown away.
- You have designed it so that parts such as motors, fastenings or packaging can be used again and it is easily taken apart.
- You have reduced the amount of waste there might be during the making of the product.

Activity
1 Ask the students to work in groups of four to six.
2 Read out the statements.
   - If the student has already incorporated this into their design, ask them to pick up a green card.
   - If they have thought about it, or are thinking about doing it, ask them to pick up a yellow card.
   - If they cannot do this for their product or design idea, ask them to pick up a red card.
3 Allow the group some time to discuss ways they can improve their design or product, eliminating the red and amber cards.

Up the wall

Aim
Developing and modelling ideas

Objective covered
To develop different strategies to elaborate, embellish, expand and develop design ideas.

Activity

⚙ This activity is helpful when all students are sorting out which ideas are worth thinking about further, e.g. after generating ideas activities (pages 34–41).

1   Ask all students to write down or sketch ideas on one sheet of A3 paper or flip chart paper for eight to ten minutes.

2   Display all the sheets by putting them on the walls around the room.

3   Give ten minutes to walk around the 'gallery' to look at other ideas and take notes.

4   Allow them to return to their original sheets, add to and refine ideas.

My designing issues

Aim
Developing and modelling ideas

Objective covered
To try alternative, sometimes unconventional, approaches for overcoming difficulties, modifying proposals and communicating these to others.

Resources
A set of cards for each group.

Possible cards

Getting the price low
Using what my research tells me
Ensuring it is safe
Making it look good
Making sure the user likes it
Making it original, different
    and better

Getting it to work well
Making it well
Finding out what the user thinks
Ensuring it is easy to use
Making it strong enough
Being able to recycle it

Activity

⚙ This activity should take place when all students are developing one or two ideas further and need advice that will help them move forward.

1   Ask the students to work in pairs or threes.

2   Give each group a set of cards with possible design issues and ask them to rank the cards in order: 'The most important design issues for me are...'

3   Leave some cards blank for them to add their own issues.

4   Ranking the issues clarifies the focus for the students and they will also start to offer solutions to the 'worries' they have.

Dice roll questions

Aim
Developing and modelling ideas

Objective covered
To continually think visually, spatially and analytically when developing and evaluating ideas.

Resources
Giant foam dice can have these questions attached to them, or else you can use ordinary dice, where the number thrown relates to the list of questions. Giant foam dice are available from Devon County Council purchasing catalogue number H 71.4950 (telephone 01392 384697).

Activity
⊛ This activity should take place when all students are developing one or two ideas further and need questions that will help them move forward.

1   Ask the students to work in pairs or threes. They should show their design ideas to the others in the group and explain them briefly.

2   Each person takes it in turn to roll the dice and think about the numbered question in relation to their design.

Hint

Numbered dice roll questions

1   Is this the best way I could do this?

2   What alternatives are there?

3   Why is this the best way of doing this?

4   Why does this idea seem feasible?

5   Am I using the best material for the job?

6   Why is it likely to work well?

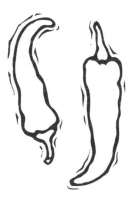

What if?

Aim
Developing and modelling ideas

Objective covered
To try alternative, sometimes unconventional, approaches for overcoming difficulties, modifying proposals and communicating these to others.

Resources
Six 'what if' cards.

Activity

⚙ This activity should take place when all students are developing one or two ideas further and need an activity that may open up different paths and more creative thinking.

1 Put six 'what if' questions on cards (see Hint).

2 Ask six of the students to pick a card at random, give them a bit of thinking time and then ask them to come to the front of the class.

3 Ask each student to talk briefly about their design and how they might change it prompted by the question.

Possible 'what if' questions

● What if ... (material) were heavily taxed due to environmental reasons?

● What if no packaging were allowed for this product, except that which was part of the design?

● What if this product were now banned for children?

● What if the same design idea as yours is produced by a competitor next week?

● What if you had to be able to use it in the outdoors as well as indoors?

● What if you had to be able to flat pack it for transport and storage?

Champions

Aim
Developing and modelling ideas

Objective covered
To draw on mathematical, scientific and technical knowledge when appraising ideas against their design criteria.

Activity

⊛ This activity should take place when students are working on a team project, when they are developing their ideas and need further suggestions.

1   Ask the students to work in groups of four.

2   For each idea in the group, find someone who is prepared to be a 'champion', who finds the idea interesting and is prepared to talk in its favour.

3   Give the students ten minutes' preparation time and five minutes' talking time each, to explain the process they worked through, the positive and negative aspects of the process itself, and the results.

Positive impact

YEAR 9

Aim
Developing and modelling ideas

Objective covered
To make decisions regarding the choice of materials and manufacturing processes and use them to draw up a manufacturing specification.

Activity
⚙ This activity should take place when all students have one or two ideas that they are developing and modifying and need ideas to help them improve.

1   List up to eight groups of people affected (good or bad) by the product during design, development, manufacture, sale, use and disposal.

2   Choose four of these groups of people and write their names in the centre of another circle, and how they will be affected around the circle.

3   Use two different highlighters to mark good points and bad points.

Hint
For further details of this 'Winners and Losers' activity, see www.stepin.org

Follow-up
Discuss with the students – where they could improve their design ideas to make sure fewer people lose out.

On the back of an envelope

Aim
Developing and modelling ideas

Objective covered
To continually think visually, spatially and analytically when developing and evaluating ideas.

Resources
Envelopes and cards. Possible questions to write on the envelopes:

- What is impossible today, but if it were possible, would change the nature of the problem forever?
- What would you do if you had all the resources (money, time, facilities) in the world to solve this problem?
- If you were superheroes with special powers and could accomplish anything, how would you handle this problem?

Activity

⊛ This activity should take place when students are working on a team project, when they have some initial ideas and need further suggestions.

1 Divide the class into groups of 4–6 students.
2 Give each team a large envelope with their question on the front.
3 Give each team five minutes to write an idea or an answer on the outside of the envelope.
4 Pass the envelope to another team. This team writes one immediate action that can be taken that lesson to work towards the ideas that's written on the outside of the envelope. This suggested action is written on a card and placed in the envelope.
5 The envelopes are passed from team to team until each team has had an opportunity to include their suggestions.
6 The envelope is returned to the originating team so that they can discuss all the suggestions.

How have we used time?

Aim
Planning

Objective covered
To log time as it is used.

Activity

1 As a class, explain the main sequence of the whole activity or project. It is helpful to have this as a visual diagram.

2 Discuss and list in order stages of designing and making, allocating a time for each stage. Put this list on the board.

3 Discuss and list tasks for their personal work, allocating time for each – including homework. Put this list on the board.

4 Ask students to copy the chart and fill in the main tasks. Encourage them to review their personal progress and to revise time allocations as they proceed.

Week	What I plan to do	How long will it take?	What I achieved	How long did it take?	How I might have used time better

Follow-up

This activity is to help students log time as it is used. The chart can also be used to:

- Check to see if a project is on track and plan actions.
- Talk about how well time was used when reviewing at the end of the project.
- Predict how much time is needed on short tasks.

For example:

- Have I done enough? If not, how can I make up?
- If I carry on as I am, will I run out of time?
- How might I do this step quicker next time?

Toolkit register

Aim
Planning

Objective covered
To prepare an ordered sequence for managing the task.

Activity

⚙ This activity is useful to develop vocabulary and to organise students at the start of a practical lesson. On the board draw five large circles. Give each one a heading for a process, for example:

1 Joining

2 Heating

3 Cutting

4 Shaping and moulding

5 Holding

As you take the register, as called each student has to quickly write the name of a tool in the right category.

Follow-up
Use this at the beginning of a practical lesson. Review the main making stages and ask the students to let you know which tools they will need to use. Leave these on the board and rub out the others. They can use this list of equipment to get everything ready at the start of the lesson.

Hint
Encourage students to name tools from across the D&T department in food/product design etc., for these generic categories. If you use this technique a lot, do the register backwards to be fair!

Personal goal

Aim
Planning

Objective covered
To share decisions about the task with teachers and/or others.

Activity
⚙ Use this activity to connect learning from the last lesson and to set a lesson target.

1 In pairs, students are asked to consider five things they learned during the last lesson and put them in an agreed order of importance.

2 Students then think about and answer the question, 'How will this information help us achieve our lesson objectives for today?'

3 Take short feedback.

4 Ask each student to set themselves a personal goal for the lesson. Ask them to write this on a sticky note with their name on and put it on the class board. The teacher can refer to these in the plenary or tick off achievements as they are completed during the lesson.

Diamond ranking

Aim
Planning

Objective covered
To decide on the main stages of making and the order in which they must be carried out.

Resources
A set of cards or paper strips with the main stages of designing and making their product.

Activity
⊛ This activity is useful to give them a whole project overview.

1 Ask the students to work in groups of three or four.

2 Ask students to put the cards in order using diamond ranking (see page 15).

Possible cards (depending on the product)
- Exploring the design brief
- Having ideas
- Collecting information about the task
- Choosing ideas
- Changing and refining ideas
- Evaluating ideas
- Testing and modelling ideas
- Modifying the product
- Planning how to make it
- Making my product
- Evaluating the project and my learning

Go with the flow

Aim
Planning

Objective covered
To reflect and evaluate how time is used within an activity.

Activity
1   Explain that a flow chart is a graphic organiser used to show a sequence of events, actions, roles or decisions.

2   Show the class an example flow chart for making a product.

3   As a pair, students make a list of the main stages of the making process.

4   Students arrange the stages in a logical sequence on the flow chart.

5   As a class, discuss how much time will be needed for each stage.

Hint
Flow charts help students sequence a series of actions or tasks chronologically. For example, students are given a selection of flow chart cards that contain the instructions to make a food product, and a selection of card symbols that link to HACCP (Hazard Analysis Critical Control Points). Students in pairs sort the flow chart cards into the correct order and then decide where to place the HACCP symbol cards.

Differentiation

Sequence mapping
⊛ This is a simple flow chart which makes use of just three shapes:
- Triangles – to indicate the start and end of a process
- Circles – to indicate actions that need to be taken
- Rectangles – to indicate events or outcomes

Go seek

Aim
Planning

Objective covered
To gather technical information about the materials and components available and use this to inform decisions about suitability for purpose.

Activity

1   As a class, brainstorm places and ways to find information to help with designing (e.g. library, Internet search, standards data).

2   Ask the students to write the main headings or categories in big letters on A4 paper/card.

3   Ask the questions below one at a time. Ask students to match them to places to find out the information by holding up the card with the answer.

Possible questions

- What is the cost of . . . (ingredient or material)?
- What is the strongest material to make a bag for hiking?
- How long does this T-shirt's sleeve need to be for an adult who normally wears a 'large'?
- Which fabrics can be used for children's nightwear?
- What's the strongest tasting cheese?
- How long can you keep this material/ingredient before it can no longer be used?
- What's the favourite colour for a camera for teenagers?
- What other products are similar to this and how are they made?
- What tool will work the best to cut this wood?

Hint
Remind students that it is important to judge whether the information source is reliable, particularly when searching the Internet.

Using a Gantt chart

Aim
Planning

Objective covered
To produce plans that allocate time needed to carry out the main stages of making.

Activity
⚙ This activity is useful to get students to take responsibility for some planning.

1 Write this blank Gantt chart on the board.
2 As a class, make a list of the main stages of the project.
3 Discuss how long each stage will take.
4 Show the class how to use the Gantt chart and how some tasks can go on at the same time.
5 Ask the students to copy the blank chart as an individual or in pairs.
   - Put the tasks in the order they need to be started.
   - Use a line to show how long each task will take.

Task	Week 1	Week 2	Week 3	Week 4	Week 5

Follow-up
Henry Laurence Gantt (1861–1919) was a mechanical engineer, management consultant and industry advisor. Henry Laurence Gantt developed Gantt charts in the second decade of the twentieth century. Gantt charts were used as a visual tool to show scheduled and actual progress of projects. For more information go to website: www.ganttchart.com

Your best bits

Aim
Evaluating

Objective covered
To make judgements about the originality and value of ideas and solutions to further their development.

Activity
* This activity can take place when the student has one idea that they are trying to develop further, modify and improve. It can also be used for evaluation once the product has been finished.

1   Ask the students to work in pairs and describe the three best features of their design idea: 'The best thing about my design is ... because ...' Allow discussion and clarification within the pairs. Prompt questions can help:

    ● Describe two things people like about your product and two things they might not like.

    ● Describe who your product is aimed at and a user test that you can carry out.

    ● Give two important points you found out about users' needs that helped you with your designing.

    ● Describe an idea that you had but have not used and explain why it was not suitable

    ● Customers may have problems using your product. Describe one possible problem that customers may have using your product and how it would be overcome.

2   Ask for feedback from a few of the students in the class, asking them to tell the rest of the class about their partner's product's best features (not their own).

ACCESS FM

Aim
Evaluating

Objective covered
To examine, describe and evaluate similar products to gain useful technical information.

Activity

⚙ This is a product analysis activity based around the easy-to-remember acronym ACCESS FM developed by Bluefish.

1 Explain to the students that this acronym represents:

Aesthetics

Cost

Customer

Environment

Safety

Size

Function

Materials

2 Ask them to work in a group of three or four students and to think of two or three similar but different products, e.g. fruit desserts, torches, storage containers, game, camera. Ask the students to use these headings to analyse products.

Hint
To make this activity shorter, you can ask the students to pick four of the eight headings of their choice or focus on one product.

How well have I done?

Aim
Evaluating

Objective covered
To evaluate their design ideas and products by comparing them against the original design criteria and suggesting improvements.

Resources
Points cards or tokens for each person. A set of four questions on cards for each group, such as:

Exploring the task
- What was the design brief?
- What was expected for my project to be a success?

Generating ideas
- What did I do to help me have ideas?
- What information did I collect to help with ideas?

Developing and modelling ideas
- What did I decide to design and make?
- Why did I choose this idea?

Making
- Did I need much help from the teacher?
- Was it easy or hard for me?

Learning
- What new things have I learned on this project?
- In what ways have I improved?

Activity
1 Ask the students to work in groups of three or four.
2 Lay four cards face down on the table.
3 Each student takes it in turn to pick a card and answer the question.
4 The rest of the group can award them up to five points for their answer.
5 All students have a turn until all the cards have been answered.

Add up the positives and negatives

Aim
Evaluating

Objective covered
To understand the need that a product is intended to serve and judge how well it meets that need.

Activity

⚙ This activity should take place when students have design ideas to evaluate.

1 Ask students to write the idea or attach a sketch to the top of an A4 page divided into two. Ask them to focus on how well they think their idea meets the needs of the user.

2 With a partner, ask students to list all the positive aspects they can think of in five minutes in one half of the page. Then list all the negative aspects they can think of in five minutes in the other half.

3 They then assign a value from 1 to 100 to every positive aspect and minus 1 to minus 100 to every negative.

4 Ask them to add up the positive and negative scores.

5 Ask them to discuss the general feelings it gives them about how well the idea meets the needs of the user and which aspects of the design idea they feel are most important (because they have been assigned a high numerical value).

Hint
The number of positives and negatives for the idea is less significant than the sum of their values.

Differentiation

⚙ You can use this technique with an existing product first, for the students to practise.

Chequerboard

Aim
Evaluating

Objective covered
To make judgements abut the originality and value of ideas and solutions to further their development.

Activity

This activity should take place when students have three or four different design ideas and want to see how far each idea meets the design criteria.

1   Ask the students to copy this chart onto a sheet of paper and to write their three or four ideas in the left-hand column or attach annotated sketches.

Idea	Criteria	Criteria	Criteria	Criteria ....
Design 1	✓			

2   Agree as a class the most important criteria for the product (you can use ACCESS FM, p. 76). Ask the students to write these in the top column.

3   In a group of three or four, the student explains each idea, and then each person in the group adds a tick when they think the idea meets that criterion. All ideas in the group are discussed in turn.

Follow-up
Ask the students if this exercise helped them to think of new ways to meet the criteria that they had not thought of themselves.

Hint
ACCESS FM can be used to focus students on the features. Criteria vary according to the product, but can include features such as:

• Easy to use   • Colourful   • Cost   • Works well   • Strong
• Safe   • Looks and feels (tastes) good

PMI

Aim
Evaluating

Objective covered
To modify and transform ideas, changing direction if needed, and ask questions (e.g. Where is this going? Does it work? What do I do next? What makes a good...?).

Activity

⊛ This activity should take place when students have three or four different design ideas and want to focus objectively on the positive and negative aspects, and react to their interesting attributes as well.

1   Ask the students to copy this chart.

Ideas	Plus	Minus	Interesting
Idea 1 .... or attach sketch			

2   Ask the students to work in pairs:

- List all the idea's positive features in the 'plus' column.
- List all the idea's negative features in the 'minus' column.
- List everything that is worth noting or commenting on, but isn't clearly a plus or a minus, in the 'interesting' column.

3   Discuss the completed chart with another pair, focusing on the interesting points.

Hint
ACCESS FM can be used to focus students on the features.

'ComPairs'

Aim
Evaluating

Objective covered
To take account of the working properties of materials when deciding how and when to use them.

Resources
Pairs of products made from different materials, e.g. bags, pens.

Activity
⊛ This activity can take place once the student has one idea that they are trying to develop further, modify and improve. It can also be used for evaluation once the product has been finished.

1   Ask the students to work in groups of four.

2   Give the students a pair of products made from different materials, e.g. a plastic bag and a paper bag, a ballpoint pen and a pen made from recycled cups.

3   Ask them to analyse each product in terms of volume, strength, stiffness, ease of packing and carrying (bags), or appearance, comfort, function (pens).

4   Take feedback from the groups.

Follow-up
Put the main heading summarising the product life cycle on the board. Ask the students to split into pairs, one pair to look at paper bags and one pair to look at plastic bags to work out the life cycle for their bag using the headings.

Make the dice roll

Aim
Evaluating

Objective covered
To identify any design weaknesses in the choice of materials and manufacturing processes.

Resources
One large die.

Activity

1   Attach these questions to a large die (or two dice), or simply number the questions 1–6 (or 1–12), and students use them according to the number thrown. Example questions for dice:

   1   Name one of the tools or items of equipment you will use, what they are for and why you chose them.

   2   Name one making technique that you will use and give simple reasons to say why it is the best one.

   3   Name one material that you will use and the main properties of it.

   4   Describe one possible risk involved in making your product and how it can be avoided.

   5   Describe one way of reducing the time taken for making your product.

   6   Describe one method of combining the materials you are using and how the performance of each material will change.

2   Ask the students to work in a group of three or four, roll the dice and answer the question in relation to their design idea.

3   At the end, ask students to spend a few minutes working on their individual designs recording changes to improve any weaknesses they've identified in the materials and making processes they have chosen.

For and against

Aim
Evaluating

Objective covered
To evaluate how their ideas and solutions would benefit individuals and the community.

Resources
Sticky name labels for the class to wear, e.g. 12 × 'In favour', 12 × 'Objects'.

Activity

1   Start the lesson with this case study. In a recent survey of residents living close to a wind farm in Cornwall, some of the residents objected to the wind farm for the following reasons:

   ● It spoils the countryside.
   ● It is noisy.
   ● It interferes with TV.
   ● It disturbs wildlife.

   Other residents were in favour of the wind farm:

   ● It is clean and does not pollute.
   ● It is safe.
   ● Wind will never run out.

2   Ask the students to work in groups of four to six.

3   Divide the groups so that equal numbers adopt the two roles (in favour/objects) to debate the issues.

4   Ask each group to discuss and draw up a list of design proposals, which will make the site better for the local residents.

5   Present their ideas to the class (adopting their roles if they want to).

Follow-up
Ask the students to think of instances where energy could be saved in their design.

How good is this product?

Aim
Evaluating

Objective covered
To formulate criteria to judge the quality of a product; the extent to which it meets the need, purpose and resource limits; and its impact on society.

Resources
A list of questions about the product. Possible questions are:

- When/where would it be used?
- How often is it used?
- Who designed and made this?
- How has it been made?
- Where is it from?
- What other products are like this?
- Who might the owner be?
- What materials and processes have been used – why?
- Does it do what it was intended to?

Activity
This activity can take place once the student has one idea that they are trying to develop. It can also be used for evaluation.

1 Ask students to work in pairs or threes.

2 Give them a set of possible questions they can ask about a set of products and allow them to choose the questions they prefer.

3 Ask the class about why they have chosen those questions.

Group crit

Aim
Evaluating

Objective covered
To review ongoing progress, Invite feedback and incorporate it into your work.

Activity

Consulting classmates can help students develop and evaluate their ideas. Designers never think of this as cheating. In the end the idea will be theirs because they have developed it further.

1 Explain that designers often sit down part-way through a project and discuss each other's work in a group crit.

2 Ask the students to work in groups of three or four.

3 Give them a few minutes to prepare to show others their work. Ask them to make sure:
- All the work is gathered together and sorted into order.
- Others can understand it.
- The best ideas are shown clearly.

4 In each group each student in turn has exactly two minutes to describe the current situation of their project. They can say:
- What they are working on.
- Why it is like it is.
- What they are pleased about.
- What they have worries about.

5 The others in the group then have three to four minutes to comment and be constructive. Each person in the group tries to give positive comments and tries to help with what they see as the weak points in the designing.

6 After the group crit each student writes down at least three action points to describe what they need to do immediately to develop their work.

Help and harm

Aim
Evaluating

Objective covered
To appreciate the conflicting demands upon designers and makers and acknowledge the balance between help and harm.

Resources
A selection of photos or real products designed to meet similar needs.

Activity

This activity can take place once the student has one idea that they are trying to develop further, modify and improve. It can also be used for evaluation once the product has been finished.

1   Students are given a selection of three or four products designed to meet similar needs, e.g. torches, fashion shoes, computer monitors, chairs.

2   Ask them to rank the products in different ways, changing the order as they go:

- Aesthetic appeal – Which product looks, tastes, feels, smells the best?

- Cost of the materials – Which product is the most expensive or best value for money?

- Functions most effectively – Which product does the best job?

- Usefulness to society – Which product helps most people? Which is most useful to the largest number of people?

3   Discuss how designers have to juggle a number of (often conflicting) criteria when they finalise their designs – compromising costs against aesthetic appeal etc.

Map it

Aim
Evaluating

Objective covered
To formulate criteria to judge the quality of a product; the extent to which it meets the need, purpose and resource limits; and its impact on society.

Activity
1   For each idea, draw a large circle on a sheet of paper.

2   Choose the most important design criteria or design features for the product.

3   Divide the circles into sections – one section for each criterion.

4   Review each idea in turn. How far does it meet the criteria?

5   Ask the students to use different colours, sizes and designs to show how important each aspect is to the whole. For example, safety may be a larger section than cost.

6   Discuss the map with a partner and answer these questions: Which idea do I think is best? Are there any criteria I want to change? Are there any missing criteria?

Follow-up
Ask students to explain why the criteria or features they have chosen are essential in a well-designed product. Ask students to combine the best aspects of their three or four design ideas into a single new proposal.

Inclusive design

Aim
Evaluating

Objective covered
To evaluate how far their ideas and solutions would benefit individuals and/or the community.

Resources
Photos or names of people on cards, such as:

- A toddler in a pushchair
- A left-handed person
- A person who does not have strong hands
- A person with a knee injury/walking stick
- A person who is partially sighted
- A person who has hearing loss
- A person with a weak back

Activity

1   Explain that sometimes we design products for a particular target group, such as a person with a specific disability. But that also we should try to make sure our designs are suitable for as wide a range of different needs as possible.

*'Design for the young and you exclude the old.
Design for the old and you include the young'*

Activity continued

2   As a class, discuss:

- Who are you designing for?
- Do they have special needs?
- Are these needs shared by others?
- Who else would benefit from this product?

3   Ask the students to work in pairs or threes. Each student will need their design idea on the table. Ask them to consider:

- What can I do to make my product more useful to more people?
- Who couldn't use this product? Does this matter?

4   Give each group a set of photos or names of people on cards (see Resources) and think about how they might need to change their design for that person to use it effectively.

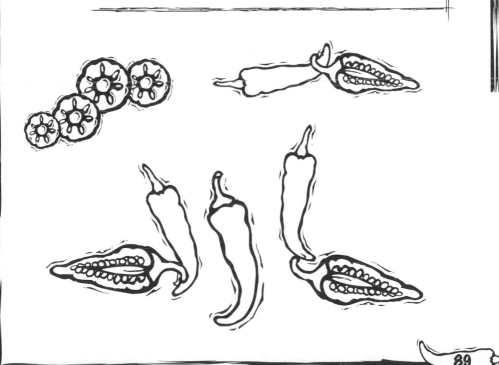

On a scale of 1-10

Aim
Evaluating

Objective covered
To review ongoing progress, invite feedback and incorporate it into work.

Activity

✿ This activity should take place as students are developing and modelling their ideas and require feedback to improve and modify them, and to reconsider the design criteria.

1   Draw this scale on the board

   0   1   2   3   4   5   6   7   8   9   10

   Then write these questions for the students:

   • Where are you now on this scale with your idea?
   • What will need to happen for your idea to get a higher score?
   • What will you need to do to make it to the top score?
   • Ask your critical friends what is right with the idea, and what still needs further work.

Follow-up
Students should record new ideas and modifications on their design, e.g. using sticky notes or annotations.

Differentiation
✿ Students can note down the ideas into categories:

   • Ideas that they can use immediately
   • Areas to explore a bit more
   • New approaches they might try

Wear a hat

Aim
Evaluating

Objective covered
To appreciate the conflicting demands upon designers and makers, and to acknowledge the balance between help and harm.

Resources
Cards with different 'hats' (viewpoints) on them. For example:

Environment
You are concerned about waste. Does it use up natural resources and energy? What about recycling, reusing, pollution, organic production?

Social
You are concerned about how it affects all groups of people. Does this product help us meet everyone's basic human needs? Does it exclude some groups?

Moral
You are concerned about whether you think this is a good product for society. Does this product improve people's lives? Is it what people want?

Economic
You are concerned with the money. How much does it cost? Do people benefit from this product? Where does the profit go?

Activity

1   Ask students to work in groups of three or four.

2   Explain that each one of them is going to evaluate the product (or proposal) from a particular viewpoint or issue.

3   Give out the 'hats' and role cards.

4   Ask the students to think about the product from the point of view on their card for five to eight minutes, making notes if they need to.

5   Then ask the group to work together, giving each 'hat' an opportunity to present their point of view about the product.

Register - safety rules

Aim
Making quality products

Objective covered
To select and use tools, equipment and processes to shape and form materials safely and accurately.

Activity
⚙ This activity is best used at the start of a practical lesson.

1 As the register is called, each student says one hygiene/safety point or rule that should be followed.

2 This involves active listening so that points are not repeated.

3 If a student cannot think of a new point, they can say 'pass' and then stand up. Give them some thinking time and return for their answer at the end of the register.

4 At the end of the register, remind the students what the most important hygiene and safety rules of this lesson are.

Follow-up
Another version of this activity would be for the first person called to say a rule and the next person to explain why that rule is important, alternating through the register.

Hint
It may be hard to think of this number of points if it is a large class, so give some clues to the students later in the register. Start from the bottom of the register at times to be fair to all students.

Acronyms help

Aim
Making quality products

Objective covered
To make single products and products in quantity, using a range of techniques to ensure consistency and accuracy.

Activity
1   Ask the students to work in groups of three or four.

2   Give each group an important principle, rule or concept written on a card.

3   Give the groups five minutes to come up with an acronym or memory technique to help people remember that rule.

4   Explain one example to give the students a clue, e.g. ACCESS FM (page 76).

Hint
Some examples are:

- Categories on the balanced plate model (page 96)
- How to set up a sewing machine or CAD machine
- Difference between thermoplastics and thermosetting plastics, hardwoods and softwoods, metals and metal alloys, frame and shell structures, a try square and an engineer's square
- What is meant by aesthetics
- Environmental issues to think about
- What is meant by target market
- What to write in a specification
- Symbols for a flow chart

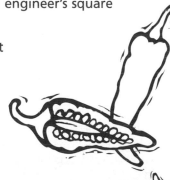

Popular plastics

Aim
Making quality products

Objective covered
To select and use tools, equipment and processes to shape and form materials safely and accurately and finish them appropriately.

Resources
A set of plastic objects, pictures or cards with names on, e.g. carrier bags, saucepan handles, flower pots, measuring jugs, food packages, crisp packets.

Activity
⊛ This activity can take place once the student has one idea that they are trying to develop further, modify and improve. It can also be used for evaluation once the product has been finished.

1   Ask the students to work in groups of three or four.

2   Ask the students to sort the objects into categories or groups. They can use a number of ways to do this – by user, function, properties (bending/strength/floats). Give them five to ten minutes.

3   Take feedback on how they grouped their plastics.

4   Ask them to spend five minutes examining one of the products to explain to the class how they think it has been made.

5   Take feedback from the groups.

Follow-up
You can discuss the properties of the plastics and how the products have been made. The students can also categorise them according to the type of plastic used. You can also do the same activity with metal products and wooden products. Possible metal products could include tools, nails, screws, nuts and bolts, kitchen equipment, jewellery, wiring, pipes, PCBs, door handles, cans, cooking foil.

Images to remember

Aim
Making quality products

Objective covered
To take account of the working characteristics and properties of
materials and components when deciding how and when to use them.

Resources
One sheet per group with a selection of signs, logos and everyday
lettering.

Activity
⊛ This activity can take place once the student has one idea
that they are trying to develop further, modify and improve.
It can also be used for evaluation once the product has been
finished.

1 Ask the students to work in pairs or threes.

2 Give each group the sheet and ask them to look at it for
no more than 30 seconds.

3 Ask them then to note and discuss in their groups the
images that had the most impact on them and why this
was.

4 Take feedback from the groups.

Follow-up
Discuss strategies to create impact and reasons for the value of shock
and impact.

Stick 'em up

Aim
Making quality products

Objective covered
To take account of the working characteristics and properties of materials and components when deciding how and when to use them.

Resources
A large blank version of the balanced plate. Cards with names or photos of foods for each group (which can be downloaded from www.nutrition.org.uk). Blu-tack.

Activity

This activity can be used to reinforce nutritional understanding and application. It can also be used for evaluation once the product has been finished.

1 Remind the class of the balanced plate model of healthy eating.

2 Ask what the sections of the chart show and annotate the headings at the side of the chart.

3 Ask the students to work in groups of four to six.

Fruit & vegetables

Bread, other cereals and potatoes

Meat, fish & alternatives

Foods containing fat
Food containing sugar

Milk & dairy foods

4 Give each group a set of photo/name cards of foods. As a food is called, one of the group comes up to the board and sticks it in the right section. (If the group are wrong they have to sit out a go.) Keep calling foods until one team has all their cards on the board.

What tools do I need?

Aim
Making quality products

Objective covered
To select and use tools, equipment and processes to shape and form materials safely and accurately and finish them appropriately.

Resources
For each group:

- A recipe or plan of work that has been cut up into steps for the main stages of making
- Cards with pictures or names of tools/equipment
- Cards with a range of 5–30 minutes
- Some blank cards

Activity
⊛ Use this activity before students get onto making.

1 Ask the students to work in pairs or threes.

2 Ask them to put the making stages in the right order.

3 Ask them to allocate the right tools and equipment cards to each stage.

4 Ask them to assign time for each step.

5 Take feedback from the groups.

Remember, remember

Aim
Making quality products

Objective covered
To learn about the working characteristics and applications of a range of modern materials.

Activity

1 Divide students into groups of three or four and ask them to invent a simple way to help them remember technical terms (rhyme, rap, mnemonic, diagram, cartoon). For example:

Forces
- Compression forces – pushing forces
- Torsion forces – twisting forces
- Shearing forces – cutting forces
- Bending forces – bending structures
- Tension forces – pulling forces

Properties
- Malleability – the ability to be worked into new shapes
- Strength – the ability to resist stress and strain
- Hardness – the ability to withstand knocks and bumps without damage
- Ductility – the ability to be stretched or squeezed into a new shape
- Weight – how heavy/light something is
- Toughness – the ability to resist breaking forces
- Conductivity – the ability to conduct heat or electricity
- Colour – the appearance and colour of the surfaces

2 Take feedback from the groups.

Differentiation
⊛ This activity can be used for any aspect of D&T.

Magical mixtures

Aim
Making quality products

Objective covered
To learn about the working characteristics and applications of a range of modern materials.

Resources
- (Custard) Electric kettle, jug and fork, pack of instant custard
- (Salad dressing) Oil, lemon juice, jar with lid
- (Meringue) Clear mixing bowl, whisk, egg white

Activity

1   Show the students three demonstrations:

- Making instant custard
- Making a salad dressing
- Making meringue

2   Before you begin each demonstration, explain what you are going to do and ask them to predict what will happen.

3   Carry out the demonstration, asking why they think it behaves in this way and how that working property is used in food designing.

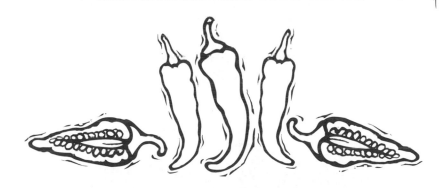

Stick it on your head

YEAR 8

Aim
Making quality products

Objective covered
To select and use tools, equipment and processes to shape and form materials safely and finish them appropriately.

Resources
Enough sticky notes for the class, with four to six key words written on them, for example:

Electronic devices	*Mechanical devices*
Resistor	DC motor
555 timer	Stepper motor
Logic gate	Relay
Capacitor	Solenoid
Diode	PCB
LED	Gearbox

Activity

⊛ This activity can take place to help revise technical terms and vocabulary, and uses for tools, equipment and processes.

1   Ask students to work in groups of four to six.

2   Each student has a mystery (unseen) sticky note attached to their forehead. On each sticky note there has been written a key word, such as the names of electronic or mechanical devices.

3   Students try to find out what their word is by asking the group suitable questions. The clue to the category of key word is given by looking at what the rest of the group are wearing.

Card pairs

Aim
Making quality products

Objective covered
To take account of the working characteristics and properties of materials and components when deciding how and when to use them.

Resources
Each group has a set of card pairs, such as:

- Fax – photocopier
- Smallpox vaccine – vitamins
- Braille – Morse code
- Sonar – radar
- Margarine – peanut butter
- Dry cleaning – blue jeans
- Helicopter – credit card
- Underground – Coca cola
- Geodesic dome – Disney
- Neon – fluorescent

Activity
⚙ Ask the students to work as a group of four to six and use the card pairs to say which invention came first. (Answer: the first one in each pair came first, so reorder them for students.)

What it says on the can!

YEAR 9

Aim
Making quality products

Objective covered
To take account of the working characteristics and properties of materials and components when deciding how and when to use them.

Resources
Product packages, textiles labels or ingredient lists from food packages. Carefully extract the information about the materials and how it is made, without revealing the actual product. Produce on OHT or photocopied sheet.

Activity
- This activity can take place once the student has one idea that they are trying to develop further, modify and improve. It can also be used for evaluation once the product has been finished.

  1 Give the students four to six sets of information from labels (see Resources).

  2 Ask them to work out what the product is and why those materials and ingredients are used in that product.

Differentiation
- You can ask the students to match the label to the product to make it easier.

Performance finishes

Aim
Making quality products

Objective covered
To select and use tools, equipment and processes to shape and form materials safely and finish them appropriately.

Resources
Cards detailing ways of finishing a product, such as:

Textiles	*Product design*
Brushing	Staining
Mothproofing	Painting
Calendaring	Varnishing
Shrink resistant	Waxing
Flame resistant	Polishing
Stain repellent	Lacquering
Anti-static	Laminating
Stiffening	
Drip dry	
Waterproofing	

Activity
⚙ This activity can take place when the student has one idea that they are trying to develop further, modify and improve. It can also be used for evaluation once the product has been finished.

1   Ask the students to work in pairs or threes.

2   Give each group a set of cards with possible ways of finishing a product.

3   Ask them to discuss what each one means and why it might be useful.

4   Then choose two or three different finishes that may be possible for their product, sketching how they will look.

Marking out, cutting and joining bingo

Aim
Making quality products

Objective covered
To take account of the working characteristics and properties of materials and components when deciding how and when to use them.

Activity

1 Ask the students to copy this chart from the board and fill in as many pieces of equipment as possible into each grid space.

2 As the register is called, ask the student to name a piece of equipment from each grid space.

3 If any student has it in the correct grid space, they can mark it off bingo-style.

4 Keep calling names and marking off until 'bingo' is called.

5 Don't forget to finish the register!

Material	Marking out	Cutting	Joining
wood			
metal			
acrylic			

Activity continued

Possible answers

Material	Marking out	Cutting	Joining
wood	Marking gauge Try square Pencil	Back saw Jigsaw Bench hook Clamp	Dowel Glue Panel pins Screw Nails Nuts and bolts Screwdriver Hammer
metal	Scriber Engineer's square	Hack saw Abrafile Bench shears Tin snips Vice	Pop rivets Pillar drill Solder
acrylic	Odd leg calipers	Coping saw Jigsaw	Glue

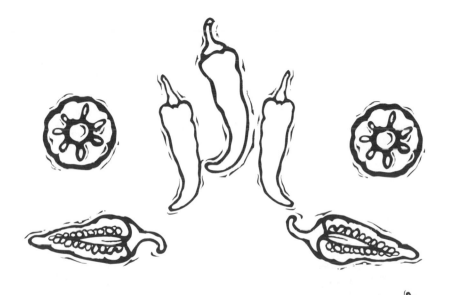

Published by Letts Educational
The Chiswick Centre
414 Chiswick High Road
London W4 5TF
☎ 020 89963333
✎ 020 87428390
✉ mail@lettsed.co.uk
🖥 www.letts-education.com

Letts Educational Limited is a division of Granada Learning Limited, part of ITV plc.

First published 2005

ISBN 1 84419 059 5

British Library Cataloguing in Publication Data
A catalogue record for this book is available from the British Library.

Commissioned by Helen Clark
Project development by Julia Swales
Editing by Nicola von Schreiber
Cover design by Ken Vail Graphic Design
Cover photo: Jan Cook/Telegraph Colour Library
Internal design by Ian Foulis & Associates
Illustrations by Ian Foulis & Associates
Typeset by FiSH Books, London
Printed and bound by Ashford Colour Press